Craters of the Moon

A Guide to Craters of the Moon
National Monument
Idaho

Produced by the
Division of Publications
Harpers Ferry Center
National Park Service

U.S. Department of the Interior
Washington, D.C.

Using This Handbook
Craters of the Moon National Monument protects
volcanic features of the Craters of the Moon lava
field. Part 1 of this handbook introduces the park
and recounts its early exploration. Part 2 explores
how life has adapted to the park's volcanic landscape—
and how people have perceived it. Part 3 presents
concise travel guide and reference materials for touring
the park and for camping.

National Park Handbooks are published to support
the National Park Service's management programs
and to promote understanding and enjoyment of the
more than 350 National Park System sites, which
represent important examples of our country's natural and cultural inheritance. Each handbook is
intended to be informative reading and a useful
guide before, during, and after a park visit. More
than 100 titles are in print. They are sold at parks
and can be purchased by mail from the Superintendent
of Documents, U.S. Government Printing Office,
Washington, DC 20402. This is handbook number 139.

Library of Congress Cataloging-in-Publication Data
Craters of the Moon: A Guide to Craters of the Moon
National Monument, Idaho/produced by the Division
of Publications, National Park Service.

p. cm.—(Official national park handbook; 139)
1. Craters of the Moon National Monument (Idaho)—
Guidebooks.
2. Geology—Idaho—Craters of the Moon National
Monument—Guidebooks. I. United States National
Park Service. Division of Publications. II. Series:
Handbook (United States. National Park Service.
Division of Publications); 139.
F752.C7C73   1991   917.96'59—dc20   89-13670CIP
ISBN 0-912627-44-1

Part 1

Welcome to Craters of the Moon

Rift Volcanism on the Snake River Plain

Light playing on cobalt blue lavas of the Blue Dragon Flows caught the inner eye of explorer Robert Limbert: "It is the play of light at sunset across this lava that charms the spectator. It becomes a twisted, wavy sea. In the moonlight its glazed surface has a silvery sheen. With changing conditions of light and air, it varies also, even while one stands and watches. It is a place of color and silence. . . ."

Limbert explored the Craters of the Moon lava field in Idaho in the 1920s and wrote those words for a 1924 issue of *National Geographic Magazine.* "For several years I had listened to stories told by fur trappers of the strange things they had seen while ranging in this region," wrote Limbert, a sometime taxidermist, tanner, and furrier from Boise, Idaho. "Some of these accounts seemed beyond belief." To Limbert it seemed extraordinary "That a region of such size and scenic peculiarity, in the heart of the great Northwest, could have remained practically unknown and unexplored. . . ." On his third and most ambitious trek, in 1920, Limbert and W. C. Cole were at times left speechless by the lava landscape they explored. Limbert recounted his impressions in magazine and newspaper articles whose publication was influential in the area's being protected under federal ownership. In 1924, part of the lava field was proclaimed as Craters of the Moon National Monument, protected under the Antiquities Act. It was created "to preserve the unusual and weird volcanic formations." The boundary has been adjusted and the park enlarged since then. In 1970, a large part of the national monument was designated by Congress as the Craters of the Moon Wilderness. It is further protected under the National Wilderness Preservation System.

Until 1986, little was known about Limbert except for those facts recounted above. That year, however, a researcher compiling a history of the national monument located Limbert's daughter in Boise. The daughter still possessed hundreds of items, including early glass plate negatives, photographs, and manuscripts of her father and that shed more light on his life, the early days of Idaho, and Craters of the Moon. Some of these photographs served as blueprints for the National Park Service in the rehabilitation of fragile spatter cone formations that have deteriorated over the years of heavy human traffic. The Limbert collection has been fully cataloged by Boise State University curators and has already proven to be a valuable resource to historians interested in Limbert and this fascinating part of Idaho. Preservation of the area owes much to Limbert's imaginative advocacy in the true spirit of the West in its earlier days.

Local legends, beginning in the late 1800s, held that this area resembled the surface of the moon, on which—it must now be remembered—no one had then walked! Geologist Harold T. Stearns first used the name Craters of the Moon when he suggested to the National Park Service, in 1923, that a national monument be established here. Stearns found "the dark craters and the cold lava, nearly destitute of vegetation" similar to "the surface of the moon as seen through a telescope." The name Craters of the Moon would stick after Limbert adopted it in *National Geographic Magazine* in 1924. Later that year the name became official when the area was set aside by President Calvin Coolidge as a national monument under the Antiquities Act.

Like some other areas in the National Park System, Craters of the Moon has lived to see the name that its early explorers affixed to it proved somewhat erroneous by subsequent events or findings. When Stearns and Limbert called this lava field Craters of the Moon, probably few persons other than science fiction buffs actually thought that human beings might one day walk on the moon and see firsthand what its surface is like. People have now walked on the moon, however, and we know that its surface does not, in fact, closely resemble this part of Idaho. Although there are some volcanic features on the surface of the moon, most of its craters were formed by the impact of meteorites colliding with the moon.

Moonscape or not, early fur trappers avoided the lava flows along the base of the Pioneer Mountains at the north of today's park. In doing so, they followed Indian trails such as one found by Limbert that "resembled a light streak winding through the lava. When the sun was directly overhead it could be seen to advantage, but at times was difficult to follow. Think of the years of travel," Limbert marveled, "necessary to make that mark on rock!" At least one Indian trail was destined to become part of Goodale's Cutoff, an alternative route on the Oregon Trail that pioneers in wagon trains used in the 1850s and 1860s. Many adjectives early used for this scene—weird, barren, exciting, awe-inspiring, monotonous, astonishing, curious, bleak, mysterious— still apply. It is not difficult today to see why pioneering folk intent on wresting a living from the land did not tackle this volcanic terrain.

Geologists possessed the proper motivation to tackle it, however. Curiosity aroused by this lava field has led several generations of geologists, beginning with Israel C. Russell in 1901 and Harold T. Stearns in the 1920s, into a deeper understanding of its volcanic origins. With ever increasing penetration of its geological history, the apparent otherworldliness of Craters of the Moon has retreated—but not entirely. The National Aeronautics and Space Administration (NASA) brought the second set of astronauts who would walk on the moon to this alien corner of the galaxy before their moonshot. Here they studied the volcanic rock and explored an unusual, harsh, and unforgiving environment before embarking on their own otherworldly adventure.

Most types of volcanic features in the park can be seen quite readily by first stopping at the visitor center and then driving the Loop Road. Far more features can be seen if you also walk the interpretive trails at the stops along the Loop Road. Still more await those who invest the time required to come to feel the mysterious timelessness and raw natural force implicit in this expansive lava field. Many travelers are en route to Yellowstone National Park and spend only a couple of hours visiting Craters of the Moon. This is ironic because here you are on the geological track of Yellowstone. In fact, Craters of the Moon represents what Yellowstone's landscape will resemble in the future, and both areas can supplement your insight into what happens when the Earth's unimaginable inner forces erupt to its surface.

Although Idaho is famous for forests, rivers, and scenic mountain wilderness, its Snake River Plain

Next two pages: *Silvery leaves of the buckwheat dot a cinder garden with such regular spacing they almost look planted. Such spacing results from the shortage of available surface water. Each plant controls with its roots the space surrounding it, discouraging competing plants. Rainwater and snowmelt penetrate volcanic cinders so readily that their moisture quickly drops beyond reach of most plants' root systems. For a close-up view of a buckwheat, see page 36.*

region boasts little of these attributes. This plain arcs across southern Idaho from the Oregon border to the Yellowstone area at the Montana-Wyoming border. It marks the trail of the passage of the Earth's crust over an unusual geologic heat source that now brings the Earth's incendiary inner workings so close to its surface near Yellowstone. This heat source fuels Yellowstone's bubbling, spewing, spouting geothermal wonders. Craters of the Moon therefore stands as a geologic prelude to Yellowstone, as its precursor and the ancestral stuff of its fiery secrets.

When did all this volcanism at Craters of the Moon happen? Will it happen again? According to Mel Kuntz and other U.S. Geological Survey geologists who have conducted extensive field research at Craters of the Moon, the volcanic activity forming the Craters of the Moon lava field probably started *only* 15,000 years ago. The last eruption in the volcanic cycle ended 2,000 years ago, about the time that Julius Ceasar ruled the Roman Empire.

Craters of the Moon is a dormant, but not extinct, volcanic area. Its sleeping volcanoes could become active again in the near future. The largest earthquake of the last quarter century in the contiguous United States shook Idaho's tallest mountain, Borah Peak, just north of here in 1983. When it did, some geologists wondered if it might initiate volcanic activity at Craters of the Moon. It did not. According to Kuntz, however, this is no reason not to expect another volcanic eruption here *soon*—probably "within the next 1,000 years." Part Two of this handbook explores the still young and rapidly evolving understanding of the fascinating geologic story of Craters of the Moon.

Today's Craters of the Moon National Monument encompasses 83 square miles of the much larger Craters of the Moon lava field. Reaching southeastward from the Pioneer Mountains, the park boundary encloses a series of fissure vents, volcanic cones, and lava flows known as the Great Rift volcanic zone. This volcanic rift zone is a line of weakness in the Earth's crust that can be traced for some 60 miles across the Snake River Plain. Recent volcanism marks much of its length. You can explore the Great Rift and some of its volcanic features via the park's 7-mile Loop Drive, as described in Part Three of this handbook. In the park's northern part you will find spatter cones, cinder cones, lava flows, lava caves, and an unexpected variety of wildflowers, shrubs, trees, and wild animals. The much larger southern part of the park, designated by Congress in 1970 as the Craters of the Moon Wilderness Area, is a vast and largely untraveled region of stark volcanic features flanking the Great Rift. It offers a challenge to serious hikers and explorers—latter day Robert Limberts—who are prepared for rugged wilderness travel.

Despite its seeming barrenness, Craters of the Moon is indeed home to a surprising diversity of plant and animal life. As Limbert noted in 1924: "In the West the term 'Lava Beds of Idaho' has always signified a region to be shunned by even the most venturesome travelers—a land supposedly barren of vegetation, destitute of water, devoid of animal life, and lacking in scenic interest.

"In reality the region has slight resemblance to its imagined aspect. Its vegetation is mostly hidden in pockets, but when found consists of pines, cedars, junipers, and sagebrush: its water is hidden deep in tanks or holes at the bottom of large 'blow-outs' and is found only by following old Indian or mountain sheep trails or by watching the flight of birds as they drop into these places to quench their thirst. The animal life consists principally of migrant birds, rock rabbits, woodchucks, black and grizzly bears: its scenery is impressive in its grandeur."

Years of patient record-keeping by scientists have fit numbers to Limbert's perceptive observations. The number of species identified includes more than 300 plants, 2,000 insects, 8 reptiles, 140 birds, 30 mammals—and one amphibian, the western toad. We now call Limbert's "rock rabbit" the pika. The grizzly is long gone here. With few exceptions, the park's denizens live mostly under conditions of great environmental stress.

Near constant winds, breeze-to-gale in strength, sweep across the park to rob moisture from all living things. Scant soils, low levels of precipitation, the inability of cinder cones to hold rainwater near the surface, and the heat of the summer sun—intensified by heat-absorbing black lavas—only aggravate such moisture theft. Cinder surfaces register summer soil temperatures of over 150°F and show a lack of plant cover. Plants cover generally less than 5 percent of

Next two pages: *Winter snow transforms these landscapes, smoothing out both contours and the jagged edges of lavas. Less lunar in appearance now, the park nonetheless maintains an otherworldly aura.*

Pages 16-17: *The park was named in 1924, 45 years before humans walked on the Moon. Although we now know more about the Moon's actual surface, the park's name still rings true. Only a few trees immediately suggest that the large photo was taken on Earth. In the inset photo, astronaut Edwin E. "Buzz" Aldrin walks on the Moon near the lunar module.*

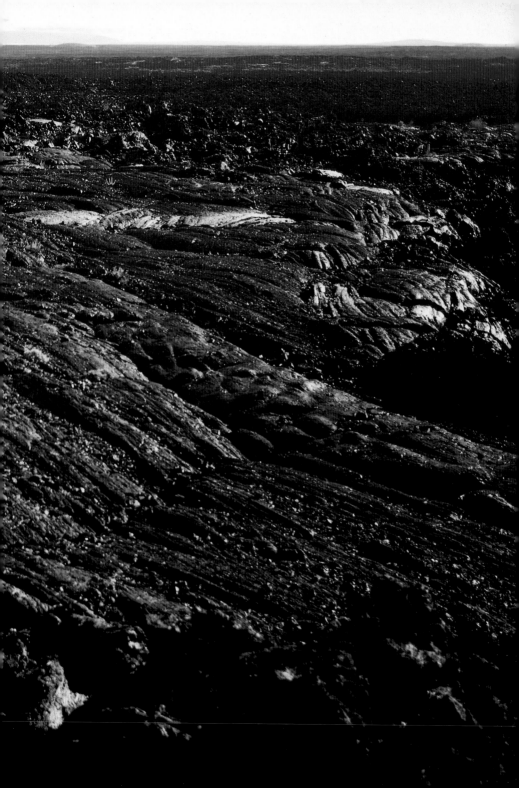

the total surface of the cinder cones. A recent study found that when the area is looked at on a parkwide basis, most of the land is very sparsely vegetated (less than 15 percent vegetative cover). On a scale of sand trap to putting green, this would approach the sand trap end of the scale.

Into this difficult environment wildlife researcher Brad Griffith ventured to count, mark, and scrutinize the mule deer of Craters of the Moon in May 1980. Griffith, of the University of Idaho, conducted a three-year study of the park's mule deer population because the National Park Service was concerned that this protected and productive herd might multiply so much that it would eventually damage its habitat. Among other things, he would find that the herd has developed a drought evasion strategy that makes it behave unlike any mule deer population known anywhere else.

"By late summer," Griffith explains, "plants have matured and dried so that they no longer provide adequate moisture to sustain the deer in this landscape that offers them no free water. Following about 12 days of warm nights and hot days in late July, the deer migrate from 5 to 10 miles north to the Pioneer Mountains. There they find free-flowing creeks and the cool, moist shade of aspen and Douglas-fir groves and wait out summer's worst heat and dryness. Early fall rains trigger the deer's return to the park's wilderness from this oasis in late September to feed on the nutritious bitterbrush until November snowfalls usher them back to their winter range."

The pristine and high-quality forage of the Craters of the Moon Wilderness Area, historically nearly untouched by domestic livestock grazing, has inspired this migratory strategy for evading drought. In effect, the mule deer make use of a dual summer range, a behavioral modification unknown elsewhere for their species.

"Their late summer and fall adaptations simply complete the mule deer's yearlong strategy for coping with the limits that this volcanic landscape imposes on them," Griffith explains.

Taking a walk in the park on a mid-summer afternoon gives you a good opportunity to experience the influence of wind, heat, and lack of moisture. The park's winds are particularly striking. The lava that has flowed out of the Great Rift has built up and raised the land surface in the park to a higher elevation than its surroundings so that it intercepts the prevailing southwesterly winds. Afternoon winds usually die down in the evening. As part of the dynamics of temperature and moisture that determine mule deer behavior, this daily wind cycle helps explain why they are more active at night than are mule deer elsewhere. These deer do not move around as much as mule deer in less ecologically trying areas. They have adapted behaviors to conserve energy and moisture in this environmentally stressful landscape.

Early mornings may find park rangers climbing up a cinder cone to count the deer, continuing the collection of data that Brad Griffith set in motion with his three-year study. The rangers still conduct spring and late summer censuses: over a recent three-year period the deer populations averaged about 420 animals. Another several years of collecting will give the National Park Service a body of data on the mule deer that is available nowhere else.

The uniqueness of this data about the park's mule deer population would surely please the booster aspect of Robert Limbert's personality. Likewise, the research challenges involved in obtaining it would appeal to his explorer self. History has justified Limbert on both counts. Publicity arising from his explorations led to creation of the national monument. Furthermore, that publicity put forth a rather heady claim that history has also unequivocally borne out: "Although almost totally unknown at present," Limbert prophesied in 1924, "this section is destined some day to attract tourists from all America. . . ."

Every year tens of thousands of travelers fulfill Robert Limbert's prophecy of more than a half-century ago.

Part 2

From Moonscape to Landscape

Geology of the Craters of the Moon

A 400-mile-long arc known as the Snake River Plain cuts a swath from 30 to 125 miles wide across southern Idaho. Idaho's official state highway map, which depicts mountains with shades of green, shows this arc as white because there is comparatively little variation here compared to most of the state. Upon this plain, immense amounts of lava from within the Earth have been deposited by volcanic activity dating back more than 14 million years. However, some of these lavas, notably those at Craters of the Moon National Monument, emerged from the Earth as recently as 2,000 years ago. Craters of the Moon contains some of the best examples of basaltic volcanism in the world. To understand what happened here, you must understand the Snake River Plain.

Basaltic and Rhyolitic Lavas. The lavas deposited on the Snake River Plain were mainly of two types classified as basaltic and rhyolitic. Magma, the molten rock material beneath the surface of the Earth, issues from a volcano as lava. The composition of this fluid rock material varies. Basaltic lavas are composed of magma originating at the boundary of the Earth's mantle and its crustal layer. Rhyolitic lavas originate from crustal material. To explain its past, geologists now divide the Snake River Plain into eastern and western units. The following geologic story relates to the eastern Snake River Plain, on which Craters of the Moon lies.

On the eastern Snake River Plain, basaltic and rhyolitic lavas formed in two different stages of volcanic activity. Younger basaltic lavas mostly lie atop older rhyolitic lavas. This portion of the plain runs from north of Twin Falls eastward to the Yellowstone area on the Wyoming-Montana border. Drilling to depths of almost 2 miles near the plain's midline, geologists found ½ mile of basaltic lava flows lying atop more than 1½ miles of rhyolitic lava flows. How much deeper the rhyolitic lavas may extend is not known. No one has drilled deeper here.

What last happened here about 2,000 years ago looked much like this photograph of a volcanic eruption in Hawaii. Bubbling, pooling, and flowing lava blanketed the landscape as molten materials poured or gushed out of the Earth. Most volcanic phenomena preserved at Craters of the Moon have been seen in action in Hawaii.

Crossing Idaho in an arc, the Snake River Plain marks the path of the Earth's crustal plate as it migrates over a heat source unusually close to the surface. It is believed that the heat source fueling Yellowstone's thermal features today is essentially the same one that produced volcanic episodes at Craters of the Moon ending about 2,000 years ago.

This combination—a thinner layer of younger basaltic lavas lying atop an older and thicker layer of rhyolitic lavas—is typical of volcanic activity associated with an unusual heat phenomenon inside the Earth that some geologists have described as a mantle plume. The mantle plume theory was developed in the early 1970s as an explanation for the creation of the Hawaiian Islands. According to the theory, uneven heating within the Earth's core allows some material in the overlying mantle to become slightly hotter than surrounding material. As its temperature increases, its density decreases. Thus it becomes relatively bouyant and rises through the cooler materials—like a tennis ball released underwater—toward the Earth's crust. When this molten material reaches the crust it eventually melts and pushes itself through the crust and it erupts onto the Earth's surface as molten lava.

The Earth's crust is made up of numerous plates that float upon an underlying mantle layer. Therefore, over time, the presence of an unusual heat source created by a mantle plume will be expressed at the Earth's surface—floating in a constant direction above it—as a line of volcanic eruptions. The Snake River Plain records the progress of the North American crustal plate—350 miles in 15 million years—over a heat source now located below Yellowstone. The Hawaiian chain of islands marks a similar line. Because the mechanisms that cause this geologic action are not well understood, many geologists refer to this simply as a heat source rather than a mantle plume.

Two Stages of Volcanism. As described above, volcanic eruptions associated with this heat source occur in two stages, rhyolitic and basaltic. As the upwelling magma from the mantle collects in a chamber as it enters the Earth's lower crust, its heat begins to melt the surrounding crustal rock. Since this rock contains a large amount of silica, it forms a thick and pasty rhyolitic magma. Rhyolitic magma is lighter than the overlying crustal rocks, therefore, it begins to rise and form a second magma chamber very close to the Earth's surface. As more and more of this gas-charged rhyolitic magma collects in this upper crustal chamber, the gas pressure builds to a point at which the magma explodes through the Earth's crust.

Explosive Rhyolitic Volcanism. Rhyolitic explosions tend to be devastating. When the gas-charged molten material reaches the surface of the Earth, the gas expands rapidly, perhaps as much as 25 to 75 times by volume. The reaction is similar to the bubbles that form in a bottle of soda pop that has been shaken. You can shake the container and the pressure-bottled liquid will retain its volume as long as the cap is tightly sealed. Release the pressure by removing the bottle cap, however, and the soft drink will spray all over the room and occupy a volume of space far larger than the bottle from which it issued. This initial vast spray is then followed by a foaming action as the less gas-charged liquid now bubbles out of the bottle.

Collectively, the numerous rhyolitic explosions that occurred on the Snake River Plain ejected hundreds of cubic miles of material into the atmosphere and onto the Earth's surface. In contrast, the eruption of Mount Saint Helens in 1980, which killed 65 people and devastated 150 square miles of forest, produced less than 1 cubic mile of ejected material. So much material was ejected in the massive rhyolitic explosions in the Snake River Plain that the Earth's surface collapsed to form huge depressions known as calderas. (Like *caldron,* whose root meaning it shares, this name implies both bowl-shaped and warmed.) Most evidence of these gigantic explosive volcanoes in the Snake River Plain has been covered by subsequent flows of basaltic lava. However, traces of rhyolitic eruptions are found along the margins of the plain and in the Yellowstone area.

Quiet Outpourings of Basaltic Lava. As this area of the Earth's crust passed over and then beyond the sub-surface heat source, the explosive volcanism of the rhyolitic stage ceased. The heat contained in the Earth's upper mantle and crust, however, remained and continued to produce upwelling magma. This was basaltic magma that, because it contained less silica than rhyolite, was very fluid.

The basalt, like the rhyolite, collected in isolated magma chambers within the crust until pressures built up to force it to the surface through various cracks and fissures. These weak spots in the Earth's crust were the results of earlier geologic activity, expansion of the magma chamber, or the formation of a rift zone.

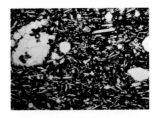

Microscopic cross sections of basaltic rock **(top)** *and rhyolitic rock show vastly different textures. Rhyolitic magma contains more silica; it is very thick and does not allow trapped gas to escape easily. Its volcanic eruptions blast large craters in the Earth's crust. Basaltic magma is more fluid and allows gas to escape readily. It erupts more gently. Here in the eastern Snake River Plain, basaltic lava flows almost completely cover earlier rhyolitic deposits.*

Identifying the Lava Flows

At Craters of the Moon the black rocks are lava flows. The surface lava rocks, basaltic in composition, formed from magma originating deep in the Earth. They are named for their appearances: **Pahoehoe** (pronounced "pah-hoy-hoy" and meaning "ropey"), **Aa** (pronounced "ah-ah" and meaning "rough"), or **Blocky.** Geologists have seen how these flows behave in modern volcanic episodes in Hawaii and elsewhere.

Pahoehoe More than half the park is covered by pahoehoe lava flows. Rivers of molten rock, they harden quickly to a relatively smooth surface, billowly, hummocky, or flat. Other pahoehoe formations resemble coiled, heavy rope or ice jams.

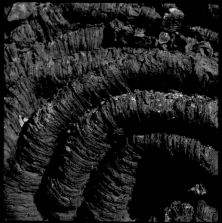
Pahoehoe lava

Aa Aa flows are far more rugged than pahoehoe flows. Most occur when a pahoehoe flow cools, thickens, and then turns into aa. Often impassable to those traveling afoot, aa flows quickly chew up hiking boots. Blocky lava is a variety of aa lava whose relatively large silica content makes it thick and often dense, glassy, and smooth.

Bombs Lava pieces blown out of craters may solidify in flight. They are classed by shape: spindle, ribbon, and breadcrust. Bombs range from ½ inch to more than 3 feet long.

Tree Molds When molten lava advances on a living forest, resulting tree molds may record impressions of charred surfaces of trees in the lava.

Aa lava

Blocky lava

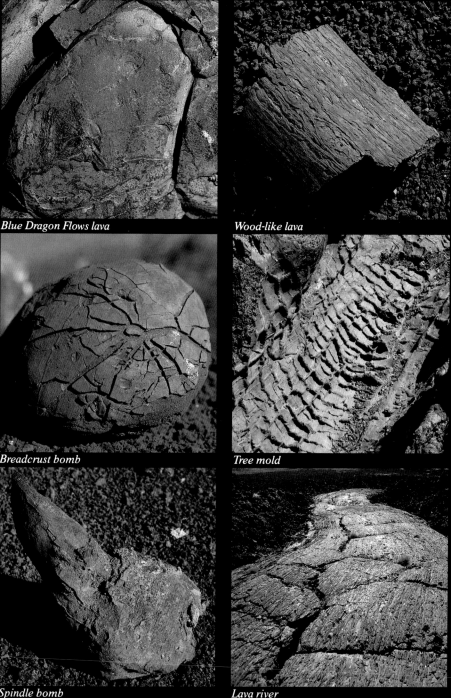

Blue Dragon Flows lava

Wood-like lava

Breadcrust bomb

Tree mold

Spindle bomb

Lava river

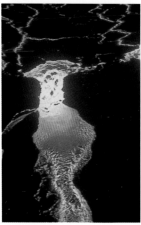

Mt. St. Helens erupts in 1980 **(top photo).** *Because the lava contained a large amount of silica, its explosive eruption contrasts sharply with recent basaltic flows in volcanic activity in Hawaii.*

Upon reaching the surface, the gases contained within the lava easily escaped and produced rather mild eruptions. Instead of exploding into the air like earlier rhyolitic activity, the more fluid basaltic lava flooded out onto the surrounding landscape. These flows were fairly extensive and often covered many square miles. After millions of years, most of the older rhyolitic deposits have been covered by these basaltic lava flows.

The Great Rift and Craters of the Moon. Craters of the Moon National Monument lies along a volcanic rift zone. Rift zones occur where the Earth's crust is being pulled in opposite directions. Geologists believe that the interactions of the Earth's crustal plates in the vicinity of the Snake River Plain have stretched, thinned, and weakened the Earth's crust so that cracks have formed both on and below the surface here. Magma under pressure can follow these cracks and fissures to the surface. While there are many volcanic rift zones throughout the Snake River Plain, the most extensive is the Great Rift that runs through Craters of the Moon. The Great Rift is approximately 60 miles long and it ranges in width from 1½ to 5 miles. It is marked by short cracks—less than 1 mile in length—and the alignment of more than 25 volcanic cinder cones. It is the site of origin for more than 60 different lava flows that make up the Craters of the Moon Lava Field.

Eight Major Eruptive Periods. Most of the lavas exposed at Craters of the Moon formed between 2,000 and 15,000 years ago in basaltic eruptions that comprise the second stage of volcanism associated with the mantle plume theory. These eight eruptive periods each lasted about 1,000 years or less and were separated by periods of relative calm that lasted for a few hundred to more than 2,000 years. These sequences of eruptions and calm periods are caused by the alternating build up and release of magmatic pressure inside the Earth. Once an eruption releases this pressure, time is required for it to build up again.

Eruptions have been dated by two methods: paleomagnetic and radiocarbon dating. Paleomagnetic dating compares the alignment of magnetic minerals within the rock of flows with past orientations of the Earth's magnetic fields. Radiocarbon dating makes use of radioactive carbon-14 in charcoal created from vegetation that is overrun by lava flows. Dates obtained by both methods are considered to be accurate to within about 100 years.

A Typical Eruption at Craters of the Moon. Research at the monument and observations of similar eruptions in Hawaii and Iceland suggest the following scenario for a typical eruption at Craters of the Moon. Various forces combine to cause a section of the Great Rift to pull apart. When the forces that tend to pull the Earth's crust apart are combined with the forces created as magma accumulates, the crust becomes weakened and cracks form. As the magma rises buoyantly within these cracks, the pressure exerted on it is reduced and the gases within the magma begin to expand. As gas continues to expand, the magma becomes frothy.

At first the lava is very fluid and charged with gas. Eruptions begin as a long line of fountains that reach heights of 1,000 feet or less and are up to a mile in length. This "curtain of fire eruption" mainly produces cinders and frothy, fluid lava. After hours or days, the expansion of gases decreases and eruptions become less violent. Segments of the fissure seal off and eruptions become smaller and more localized. Cinders thrown up in the air now build piles around individual vents and form cinder cones.

With further reductions in the gas content of the magma, the volcanic activity again changes. Huge outpourings of lava are pumped out of the various fissures or the vents of cinder cones and form lava flows. Lava flows may form over periods of months or possibly a few years. Long-term eruptions of lava flows from a single vent become the source of most of the material produced during a sustained eruption. As gas pressure falls and magma is depleted, flows subside. Finally, all activity stops.

When Will the Next Eruption Occur? Craters of the Moon is not an extinct volcanic area. It is merely in a dormant stage of its eruptive sequence. By dating the lava flow, geologists have shown that the volcanic activity along the Great Rift has been persistent over the last 15,000 years, occuring approximately every 2,000 years. Because the last eruptions took place about 2,000 years ago, geologists believe that eruptions are due here again—probably within the next 1,000 years.

PACIFIC

Pacific Plate Motion

Hawaiian Islands

OCEAN

From the air the Great Rift looks like an irregularly dashed line punctuated by telltale cones and craters **(top).** *Chainlike, the Hawaiian group of islands traces the migration of Earth's crustal plate over an unusual undersea heat source. The Hawaiian chain of islands and the Snake River Plain map similar happenings.*

Indian Tunnel

Indian Tunnel looks like a cave, but it is a lava tube. When a pahoehoe lava flow is exposed to the air, its surface begins to cool and harden. A crust or skin develops. As the flow moves away from its source, the crust thickens and forms an insulating barrier between cool air and molten material in the flow's interior. A rigid roof now exists over the stream of lava whose molten core moves forward at a steady pace. As the flow of lava from the source vent is depleted, the level of lava within the molten core gradually begins to drop. The flowing interior then pulls away from the hardening roof above and slowly drains away and out. The roof and last remnants of the lava river inside it cool and harden, leaving a tube.

Many lava tubes make up the Indian Tunnel Lava Tube System. These tubes formed during the same eruption within a single lava flow whose source was a fissure or crack in the Big Craters/Spatter Cones area. A tremendous amount of lava was pumped out here, forming the Blue Dragon Flows. (Hundreds of tiny crystals on its surface produce the color blue when light strikes them.) Lava forced through the roof of the tube system formed huge ponds whose surfaces

Lava tube

Great horned owl

cooled and began to harden. Later these ponds collapsed as lava drained back into the lava tubes. Big Sink is the largest of these collapses. Blue Dragon Flows cover an area of more than 100 square miles. Hidden beneath are miles of lava tubes, but collapsed roof sections called skylights provide entry to only a small part of the system. Only time, with the col-lapse of more roofs, will reveal the total extent of the system.

Stalactites Dripped from hot ceilings, lava forms stalactites that hang from above. **Mineral deposits** Sulfate compounds formed on many lava tube ceilings from volcanic gases or by evaporation of matter leached from rocks above. **Ice** In spring, ice stalactites form on cave ceilings and walls. Ice stalagmites form on the cave floor. Summer heat destroys these features. **Wildlife** Lava tube beetles, bushy-tailed woodrats (packrats), and bats live in some dark caves. Violet-green swallows, great horned owls, and ravens may use wall cracks and shelves of well-lit caves for nesting sites.

Icicles (ice stalactites)

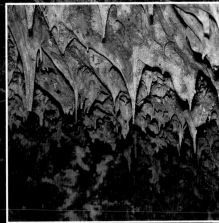

Lava stalactites

Cinder Cones and Spatter Cones

Cinder Cones When volcanic eruptions of fairly moderate strength throw cinders into the air, cinder cones may be built up. These cone-shaped hills are usually truncated, looking as though their tops were sliced off. Usually, a bowl- or funnel-shaped crater will form inside the cone. Cinders, which cooled rapidly while falling through the air, are highly porous with gas vesicles, like bubbles. Cinder cones hun-

dreds of feet high may be built in a few days. Big Cinder Butte is a cinder cone. At 700 feet high it is the tallest cone in the park. The shape develops because the largest fragments, and in fact most of the fragments, fall closest to the vent. The angle of slope is usually about 30 degrees. Some cinder cones, such as North Crater, the Watchman, and Sheep Trail Butte, were built by more than one eruptive episode.

Younger lava was added to them as a vent was rejuvenated. If strong winds prevailed during a cinder cone's formation, the cone may be elongated—in the direction the wind was blowing—rather than circular. Grassy, Paisley, Sunset, and Inferno Cones are elongated to the east because the dominant winds in this area come from the west. The northernmost section of the Great Rift contains the most

cinder cones for three reasons: **1.** There were more eruptions at that end of the rift. **2.** The lavas erupted there were thicker, resulting in more explosive eruptions. (They are more viscous because they contain more silica.) **3.** Large amounts of groundwater may have been present at the northern boundary of the lavas and when it came in contact with magma it generated huge amounts of steam. All of these conditions lead to more extensive and more explosive eruptions that tend to create cinder cones rather than lava flows.

Spatter Cones When most of its gas content has dissipated, lava becomes less frothy and more tacky. Then it is tossed out of the vent as globs or clots of lava paste called spatter. The clots partially weld together to build up spatter cones. Spatter cones are typically much smaller than cinder cones, but they may have steeper sides. The Spatter Cones area of the park (Stop 5 on the map of the Loop Drive) contains one of the most perfect spatter-cone chains in the world. These cones are all less than 50 feet high and less than 100 feet in diameter.

Life Adapts to a Volcanic Landscape

Two thousand years after volcanic eruptions subsided, plants and animals still struggle to gain toeholds on this unforgiving lava field. Much of the world's vegetation could not survive here at all. Environmental stresses created by scant soil and minimal moisture are compounded by highly porous cinders that are incapable of holding water near the ground surface where plants and other organisms can make ready use of it. Scarce at best—total average precipitation is between 15 to 20 inches per year—rainwater and snowmelt quickly slip down out of reach of the plants growing on cinder cones. Summer's hot, dry winds rob moisture from all living things exposed to them. Whisking across leaves and needles the winds carry away moisture precious to plant tissues. On the side of a cinder cone, summer day temperatures at ground level can be more than 150°F.

The secret to survival here is adaptation. Most life forms cope by strategies of either resisting or evading the extremes of this semi-arid climate. To resist being robbed of moisture by winds and heat, a plant may feature very small leaves that minimize moisture loss. To evade heat, wind, and aridity, another plant may grow inside a crevice that provides life-giving shade and collects precious moisture and soil particles. Another plant may spend about 95 percent of the year dormant. It may rush through the germination, sprouting, leafing out, blooming, and fruiting stages and return to the dormancy of its seed stage in just two weeks. The dwarf buckwheat has adapted to life on porous cinders by evolving a root system that may spread out for up to 3 feet to support its aboveground part, which is a mere 4 inches high. This buckwheat only looks like a dwarf because you can not see its roots.

Ecological conditions at Craters of the Moon are generally so harsh that slight changes can make the difference for the survival of a plant or other organism. Life thrives in many rock crevices that are surrounded

Lichens often pioneer new life on Earth. Two plants in one, lichens are composed of an alga and a fungus growing together to their mutual benefit, usually on rock. Hardy and slow-growing, lichens help break down rock to soil-building mineral matter.

Eventually their vegetable matter decays, helping to form the first soils that other plants can then use. Tough in the extreme, some lichens can be heated to high temperatures and still be capable of resuming normal growth when returned to viable conditions.

Plants Adapt to a Volcanic Landscape

Water is the limiting factor in plant growth and reproduction both on the lava fields of Craters of the Moon and on the surrounding sagebrush steppe. Plants have developed a combination of adaptations to cope with drought conditions. There are three major strategies:

1. Drought tolerance Physiological adaptations leading to drought tolerance are typical of desert plant species. The tissues of some plants can withstand extreme dehydration without suffering permanent cell damage. Some plants can extract water from very dry soils. Sagebrush and antelope bitterbrush exemplify drought tolerance.

2. Drought avoidance Certain structural modifications can enable plants to retain or conserve water. Common adaptations of this type include small leaves, hairiness, and succulence. The small leaves of the antelope bitterbrush expose less area to evaporative influences such as heat and wind.

Dwarf monkeyflower

Buckwheat

Hairs on the scorpionweed reduce surface evaporation by inhibiting air flow and reflecting sunlight. Succulent plants such as pricklypear cactus have tissues that can store water for use during drought periods. Other plants, such as wire lettuce, avoid drought by having very little leaf surface compared to their overall volume.

3. Drought escape Some plants, such as mosses and ferns, escape drought by growing near persistent water supplies such as natural potholes and seeps from ice caves. Many other drought escapers, such as dwarf monkeyflower, simply carry out their full life cycle during the moist time of the year. The rest of the year they survive in seed form.

Lava flows Most plants cannot grow on lava flows until enough soil has accumulated to support them. The park's older volcanic landscapes, where soils are best developed, are clothed with sagebrush-grassland vegetation. On younger lava flows, bits of soil first accumulate in cracks, joints, and crevices. It is in these microhabitats that vascular plants may gain footholds. Narrow cracks and joints may contain desert parsley and lava phlox. Shallow crevices will hold scabland penstemon, fernleaf fleabane, and gland cinquefoil. Deep crevices can support the syringa, various ferns, bush rockspirea, tansy-bush, and even limber pine. Not until full soil cover is achieved can the antelope bitterbrush, rubber rabbitbrush, and sagebrush find suitable niches. On lava flows soils first form from eroded lava and the slow decomposition of lichens and other plants able to colo-nize bare rock. These soils can be supplemented by wind-blown soil particles until vascular plants gain footholds. As plants begin to grow and then die, their gradual decomposition adds further soil matter. These soil beginnings accumulate in cracks and crevices, which also provide critical shade and wind protection. Deep crevices provide lower temperatures favoring plant survival.

Cinder gardens Compared to the lava flows, cinder cones are much more quickly invaded by plants. Here, too, however, volcanic origins influence plant growth. Compared to the relatively level lava flows, steeply sloping cinder cones introduce a new factor that controls the development of plant communities: topography. Here you find marked differences in the plant communities between the north- and south-facing slopes. South-facing slopes are exposed to prolonged, intense sunlight, resulting in high evaporation of water. Because of the prevailing winds, snow accumulates on northeast sides of cones, giving them far more annual water than southwest-facing sides receive. The pioneering herbs that first colonize cinder cones will persist on southwest-facing slopes long after succeeding plant communities have come to dominate north-facing slopes. It is on these north-facing slopes that limber pine first develops in the cinder garden. South-facing slopes may never support the limber pine but may be dominated by shrubs. Unweathered cinder particles range in size from 3 to 4 inches in diameter down to very small particles. They average about ¼ inch in diameter.

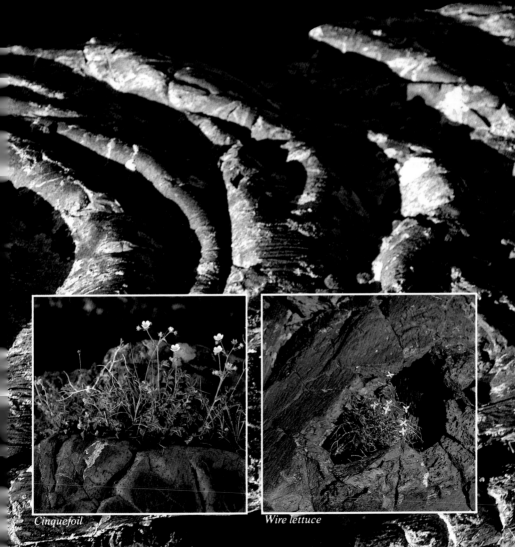

Cinquefoil

Wire lettuce

Limber pines are the tree pioneers of the lava terrain. Their seedlings (top) often find suitable conditions for germination in rock crevices long before surrounding landscapes support tree growth. Most common of all the park's trees, limber pine is named for its flexible branches. Many park animals depend on this tree in some fashion for their livelihoods. Limber pine cones stay green and resinous through their first year of development and then turn brown and woody as their seeds mature in the second year. Cones grow to about 4 inches long.

by barren exposed lava rock of the same physical composition. These microhabitats provide the critical shade and increased soil and moisture content required for plant survival. Over the years, particles of soil will naturally collect in rock crevices, which also have the effect of funneling precipitation into their depths. Their shade further protects these pockets of soil and water from wind erosion, excessive heat, and evaporation and leaching by direct sunlight.

At Craters of the Moon, crevices are of such importance to plants that botanists differentiate between narrow, shallow, and deep crevices when studying this phenomenon. Narrow crevices will support dwarf goldenweed or hairy goldaster. Shallow crevices support scabland penstemon, fernleaf fleabane, and gland cinquefoil. Deep crevices give rise to syringa, ferns, bush rockspirea, tansybush, Lewis mockorange, and even the limber pine tree. Complete soil cover and then vegetative cover can develop on these lava flows only after crevices have first become filled with soil.

Plants exploit other means of protection to survive in this harsh environment. Shaded and wind-sheltered, the northern side of a cinder cone can support grass, shrubs, and limber pine trees while the cone's southern face supports only scattered herbs. Most cinder cones in the park show distinct differences of plant cover between their northern and southern exposures. Northern exposures are cooler and more moist than southern exposures, which receive far more direct sunlight. In addition, here at Craters of the Moon, the prevailing southwesterly winds compound the ability of the dry heat to rob porous cinder cone surfaces and their living organisms of precious moisture.

The build-up of successive lava flows has so raised the landscape that it now intercepts wind currents that operate higher above surrounding plains. Limber pine trees find footholds on the shaded and sheltered northern exposures of cinder cones. Bitterbrush and rabbitbrush shrubs that can barely survive on the lower skirts of a cinder cone's southern side may grow two-thirds of the way up its protected northern face. For many species of plants the limits of habitability on this volcanic landscape are narrowly defined. Very small variations in their situations can determine success or failure.

Travelers often ask park rangers whether or not some of the park's plants were planted by people. The plants in question are dwarf buckwheats and grow in cinder gardens. It is their incredibly even spacing that creates an orderliness that is easy to mistake for human design. The regular spacing comes about because of the competition for moisture, however. The root systems of these plants exploit the available water from an area of ground surface much larger than the spread of their foliage. In this way, mature plants can fend off competition by using the moisture that would be required for a potentially encroaching plant to become established. The effect is an even spacing that makes it appear, indeed, as though someone had set out the plants on measured centers.

Craters of the Moon abounds with these surprising plant microhabitats that delight explorers on foot. The bleak lava flows separate these emerging pockets of new life, isolating them like islands or oases within their barren volcanic surroundings.

Scientists have studied Carey Kipuka, an island of plantlife in the most southern part of the park, to find out what changes have occurred in the biologic community. *Kipuka* is a Hawaiian name given to an area of older land that is surrounded by younger lava flows. Recent lava flows did not overrun Carey Kipuka, so its plant cover is unaltered. Shortage of water protected it from livestock grazing that might have changed its character. Its vegetation is a benchmark for comparing plant cover changes on similar sites throughout southern Idaho.

For the National Park Service and other managers of wildlands, kipukas—representing isolated and pristine plant habitat unchanged by human influence—provide the best answer that we have to the important question, "What is natural?" Armed with a satisfactory answer to that question, it is possible to manage the land ecologically. Park managers can seek to restore natural systems and to allow them to be as self-regulating as possible. It is ironic that Craters of the Moon, a volcanic landscape subjected to profound change, should also protect this informative glimpse of what remains unchanged.

From the park's mazes of jumbled rock, ground squirrels fashion homes with many entrances and exits. Opportunistic feeders on vegetable matter, these engaging rodents fall prey to hawks and owls from above and small predatory mammals on the ground. They therefore serve as an important transfer point between plant and animal layers of the park's food energy scheme. In the 1920s, members of the Limbert Expedition, described on pages 50 and 51, followed the flight of doves (bottom) to locate water as they explored what later became the park.

Wildflowers

Wildflowers carpet Craters of the Moon's seemingly barren lava fields from early May to late September. The most spectacular shows of wildflowers come with periods of precipitation. In late spring, moisture from snowmelt—supplemented now and then by rainfall—sees the blossoming of most of the delicate annual plants.

Many of the park's flowering plants, having no mechanisms for conserving moisture, simply complete their life cycles before the middle of summer. This is particularly true of those that grow on the porous cinder gardens into which moisture quickly descends beyond reach of most plants' root systems.

As summer continues and supplies of moisture slowly dwindle, only the most drought-resistant of flowering plants continue to grow and to bloom. With the onset of autumn rains, only the tiny yellow blossoms of the sagebrush and rabbitbrush remain.

Blazing star

Monkeyflower

Desert parsley

Wild onion

Scabland penstemon

Bitterroot

Arrow-leaved balsamroot

Paintbrush

Scorpionweed

Mule Deer

Brad Griffith could be called the mule deer man. In 1980, this wildlife researcher began a three-year study of the mule deer herd that summers in the park. The immediate concern was that the deer, protected inside the park, might be overpopulating their range and endangering their habitat. Griffith set out to find out just how the deer use the area, what their population level is, and how certain factors— production, mortality, and distribution—affect their population dynamics. The mule deer use the park April through November only, because winter brings snows too deep for the deer to find food here. The most striking finding of Griffith's research is that the mule deer at Craters of the Moon—unlike mule deer studied elsewhere—have a dual summer range. Put simply, the mule deer have had to undergo behavior modification to live here. The deer move back into the southern park in mid-April, living in the protected wilderness area there. While in the wilderness area, the park's deer routinely live up to nearly 10 miles from open water, getting their water from food, dew, fog, and temporary puddles. This area has higher quality forage for these deer than any other part of their annual range. The trade-off is that the wilderness area has almost no open water. When the moisture content of their forage decreases in summer, usually in July, the deer move up to the northern part of the park

where there is open water. Their habits in the northern part of the park are unusual, too, Griffith says, because there the deer live in much closer quarters than other herds are known to tolerate on summer ranges. They live in this wildlife equivalent of an apartment complex until the fall rains come. Then they move back down to the wilderness area. The deer make this unusual summer migration, Griffith suggests, to avail themselves of the high quality forage in the southern park. "The park serves as an island of high quality habitat for mule deer," he wrote in his report. It is now known the deer will leave the wilderness area for the northern park after 12 days with daytime highs above 80°F and nighttime lows above 50°F in summer. "We can't really predict this," Park Ranger Neil King says, "but the deer know when this is." What is happening is that the percentage of water in their forage plants falls below what is necessary to with increasi As you would nursing two fa ple days earlie only one fawn which their fa the fall of the ing. "This is ar ductive herd, "right up there est fawn survi western mule rangers contir studies by tak counts.

Indians, Early Explorers
And Practicing Astronauts

Not surprisingly, archeologists have concluded that Indians did not make their homes on this immense lava field. Astronauts would one day trek about Craters of the Moon in hopes that experiencing its harshly alien environment would make walking on the moon less disorienting for them. No wonder people have not chosen to live on these hot, black, sometimes sharp lava flows on which you must line the flight of doves to locate drinking water.

Indians did traverse this area on annual summer migrations, however, as shown by the developed trails and many sites where artifacts of Northern Shoshone culture have been found. Most of these archeological sites are not easily discerned by the untrained eye, but the stone windbreaks at Indian Tunnel are easily examined. Rings of rocks that may have been used for temporary shelter, hunting blinds, or religious purposes, numerous stone tools, and the hammerstones and chippings of arrowhead making are found scattered throughout the lava flows. Some of the harder, dense volcanic materials found here were made into crude cutting and scraping tools and projectile points. Such evidence suggests only short forays into the lavas for hunting or collecting by small groups.

The Northern Shoshone were a hunting and gathering culture directly dependent on what the land offered. They turned what they could of this volcanic environment to their benefit. Before settlement by Europeans, the vicinity of the park boasted several game species that are rare or absent from Craters of the Moon today. These included elk, wolf, bison, grizzly and black bear, and the cougar. Bighorn sheep, whose males sport characteristic headgear of large, curled horns, have been absent from the park since about 1920.

Military explorer U.S. Army Capt. B.L.E. Bonneville left impressions of the Craters of the Moon lava field in his travel diaries in the early 1800s. In *The Adventures of Captain Bonneville*, which were based

The Northern Shoshone regularly passed through the Craters of the Moon area on their annual summer migration from the Snake River to the Camas Prairie, west of the park. They took this journey to get out of the hot desert and into the cooler mountains. There they could gather root crops and hunt marmots, jackrabbits, porcupines, and ground squirrels. As they passed through today's park, they left behind arrowheads, choppers, and scrapers and built stone circles (see photo) *that may have been used for ceremonial purposes. These artifacts and structures are evidence the Indians were temporary visitors to this vast volcanic landscape.*

on the diaries, 19th-century author Washington Irving pictures a place "where nothing meets the eye but a desolate and awful waste, where no grass grows nor water runs, and where nothing is to be seen but lava." Irving is perhaps most famous for *The Legend of Sleepy Hollow,* but his *Adventures* is considered a significant period work about the West and provided this early, if brief, glimpse of a then unnamed Craters of the Moon.

Pioneers working westward in the 19th century sought either gold or affordable farm or ranch lands so they, like the Northern Shoshone, bypassed these lava wastes. Later, nearby settlers would venture into this area in search of additional grazing lands. Finding none, they left Craters of the Moon substantially alone.

Early pioneers who left traces in the vicinity of the park did so by following what eventually came to be known as Goodale's Cutoff. The route was based on Indian trails that skirted the lava fields in the northern section of the park. It came into use in the early 1850s as an alternate to the regular route of the Oregon Trail. Shoshone Indian hostilities along the Snake River part of the trail—one such incident is memorialized in Idaho's Massacre Rocks State Park—led the emigrants to search for a safer route. They were headed for Oregon, particularly the Walla Walla area around Whitman Mission, family groups in search of agricultural lands for settlement. Emigrants traveling it in 1854 noticed names carved in rocks and trees along its route. It was named in 1862 by travelers apparently grateful to their guide, Tim Goodale, whose presence, they felt, had prevented Indian attacks. Illinois-born Goodale was cut in the mold of the typical early trapper and trader of the Far West. He was known to the famous fur trade brothers Solomon and William Sublette. His name turned up at such fur trade locales as Pueblo, Taos, Fort Bridger, and Fort Laramie over a period of at least 20 years.

After the discovery of gold in Idaho's Salmon River country, a party of emigrants persuaded Goodale to guide them over the route they would name for him. Goodale was an experienced guide: in 1861, he had served in that capacity for a military survey west of Denver. The large band of emigrants set out in July and was joined by more wagons at Craters of the

Moon. Eventually their numbers included 795 men and 300 women and children. Indian attacks occurred frequently along the Oregon Trail at that time, but the size of this group evidently discouraged such incursions. The trip was not without incident, but Goodale's reputation remained sufficiently intact for his clients to affix his name to the route. Subsequent modifications and the addition of a ferry crossing on the Snake River made Goodale's Cutoff into a popular route for western emigration. Traces of it are still visible in the vicinity of the park today.

Curiosity about this uninhabitable area eventually led to more detailed knowledge of Craters of the Moon and knowledge led to its preservation. Geologists Israel C. Russell and Harold T. Stearns of the U.S. Geological Survey explored here in 1901 and 1923, respectively. Taxidermist-turned-lecturer Robert Limbert explored the area in the early 1920s. Limbert made three trips. On the first two, he more or less retraced the steps of these geologists. On his third and most ambitious trek, Limbert and W. L. Cole traversed what is now the park and the Craters of the Moon Wilderness Area south to north, starting from the nearby community of Minidoka. Their route took them by Two Point Butte, Echo Crater, Big Craters, North Crater Flow and out to the Old Arco-Carey Road, then known as the Yellowstone Park and Lincoln Highway. These explorations and their attendent publicity in *National Geographic Magazine* were instrumental in the proclamation of Craters of the Moon as a national monument by President Calvin Coolidge in 1924.

Since Limbert's day, astronauts have walked both here and on the moon. Despite our now detailed knowledge of the differences between these two places, the name—and much of the park's awe-inspiring appeal—remains the same. It is as though by learning more about both these niches in our universe we somehow have learned more about ourselves as well.

STATE
OF
IDAHO
(mid-1800s)

SHOSHONE-
BANNOCK   Craters of the Moon

Goodale's Cutoff   Trail

Snake   River   Oregon   Trail

TERRITORY

In the mid-1800s the Oregon Trail served as a major route to the West for pioneers. But when hostilities developed along the trail with the Shoshone-Bannock Indians, many of the emigrants began using an alternate route known as Goodale's Cutoff. This trail went further north and passed through the present-day park boundary.

Early Explorers and the Limbert Expedition

The first known explorations of these lava fields were conducted by two Arco, Idaho, cattlemen in 1879. Arthur Ferris and J.W. Powell were looking for water for their livestock. The first scientific explorations were carried out by Israel C. Russell, surveying the area for the U.S. Geological Survey in 1901 and 1903. Beginning in 1910, Samuel A. Paisely, later to become the park's first custodian, also explored these lava fields. In 1921, the U.S.G.S. sent two geologists here, Harold T. Stearns and O.E. Meinzer, with a geologist from the Carnegie Institute. Based on this field work, Stearns recommended that a national monument be created here. Also during the early 20s, the explorations of Idaho entrepreneur Robert W. Limbert caught the public's fancy. A report of the explorations of "Two-gun" Bob Limbert was published in the March 1924 *National Geographic Magazine*. Limbert was a Boise, Idaho, taxidermist, tanner, and furrier. He wa[s...] teur wrestler a[nd...] artist who late[r...] the national le[...] Reportedly, Li[m]-lenged Al Cap[one...] duel at 10 pac[es...] Capone declin[ed...] made three tr[eks...] fields betwee[n...] He first explo[red...] ily accessible [...] of the lava fie[lds...] third expediti[on...] area from sou[th...] ever, starting [...]

The Limbert Trek

On his third expedition, Limbert, Cole, and a dog traversed the lava flows from south to north. The photos that appeared in The National Geographic Magazine *in 1924 were taken on various expeditions.*

With Limbert were W.L. Cole and an Airedale terrier. Taking the dog along was a mistake, Limbert wrote, "for after three days' travel his feet were worn raw and bleeding." Limbert said it was pitiful to watch the dog as it hobbled after them. The landscape was so unusual that Limbert and Cole had difficulty estimating distances. Things would be half again as far away as they had reckoned. In some areas their compass needles went wild with magnetic distortions caused by high concentrations of iron in the lava rock. Bizarre features they found—such as multicolored, blow-out craters—moved Limbert to write: "I noticed that at places like these we had almost nothing to say." Limbert and Cole discovered ice caves with ice stalactites. They found water by tracking the flights of mourning doves. They found pockets of cold water (trapped above ground by ice deposits below the surface) covered with yellowjackets fatally numbed by the cold. They drank the water anyway. In desert country, said Limbert, one can't be too picky. Between Limbert's lively article in the *National Geographic Magazine,* and the reports of geologist Stearns, President Calvin Coolidge was induced to designate part of the lava fields as Craters of the Moon National Monument on May 2, 1924.

Part 3

Guide and Adviser

Approaching Craters of the Moon

Craters of the Moon National Monument is located in south-central Idaho's Snake River Plain, midway between Boise, Idaho, and Grand Teton and Yellowstone National Parks. The park includes 53,545 acres, and the elevation at the visitor center is 5,900 feet above sea level. U.S. 20-26-93 gives access to the park. Nominal entrance fees are charged. Golden Eagle, Golden Age, and Golden Access passports are honored and may be obtained at the entrance station.

Seasons and Weather. Park facilities are open and naturalist programs are conducted from mid-June through Labor Day. From November to April, the Loop Drive (see map) is closed by snow and park facilities are limited. In spring and fall, the opening and closing of facilities and the Loop Drive are determined by weather conditions, which vary greatly from year to year. In spring the weather is unpredictable. Strong winds may occur and snow and/or freezing rain are not uncommon. Temperatures range from highs in the 50s to lows in the 20s°F. Summer features warm to hot days and cool nights. Expect afternoon winds. There may be very sporadic afternoon thunderstorms, and temperatures may range from the 40s to the 90s. Fall offers generally fair weather with low precipitation and infrequent winds. Early snowstorms are possible, and snow is certain by late fall. Fall temperatures range from highs in the 60s to lows in the 30s. Winter brings the possibility of severe storms with drifting snow. Highway access is often best described as snowpacked. On bright sunny days temperatures may reach into the 40s, but the range is generally from highs in the 30s to lows around minus 10.

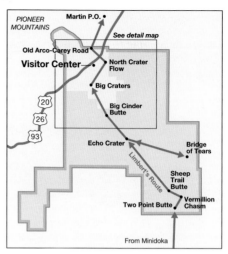

Today's park encompasses a small portion of the Great Rift and the greater portion of the Craters of the Moon Lava Field near Arco, Idaho. Blue arrows on this map show the route of the Limbert Expeditions in the 1920s. The detail map appears on page 58.

Handicapped Access. The park visitor center, restrooms, and amphitheater are accessible to the disabled.

Travel Planning. U.S. 20-26-93 provides access to Craters of the Moon. No public transportation serves the park. Scheduled airlines serve Idaho Falls, Twin Falls, and Hailey, Idaho. Rental cars are generally available at these airports, but advance reservations are advised. It is about a three-hour drive from the park to Grand Teton National Park, and about a four-hour drive to Yellowstone National Park. The official Idaho State Highway Map is available from the Idaho Transportation Department, P.O. Box 7129, Boise, ID 83707, telephone (208) 334-8000. Idaho's travel office provides information about cultural activities, scenic tours, outfitters and guides, chambers of commerce, hotels, and motels throughout the state. Write or call Idaho's Travel Promotion Division, 700 West State Street, Boise, ID 83720-2700, telephone (208) 334-2470.

Stay on Roads. Please stay on roadways and parking pullouts that are provided. If a vehicle goes off the roadway onto cinders, the cinders are compacted and the tracks may remain visible for 10 to 20 years or more.

Information about the Park. Address specific inquiries about the park to the **Superintendent, Craters of the Moon National Monument, P.O. Box 29, Arco, ID 83213,** or telephone (208) 527-3257.

Make the visitor center **(top)** *your first stop in the park. Ask at the information desk for schedules of ranger-led walks, talks, and other programs and for advice about camping.*

Photo page 57: *Visitors read a wayside exhibit beneath imposing monoliths. Flows of lava rafted such fragments of broken crater walls into otherwise inexplicable positions.*

Visitor Center and Programs

The park visitor center is located near the only entrance to the park. Here you will find displays and information to help you plan your visit. Slides, postcards, maps, and other publications about the park are displayed for sale. Park Service rangers at the information counter can answer your questions and help you plan your stay in the park.

The displays alert you to wildflowers and wild animals you might expect to see here. Other exhibits describe the park's geologic history. A film explains how lava flowed from fissures in the Earth to create the cinder cones, lava flows, and other volcanic features you will see at Craters of the Moon. The film includes actual footage of eruptions of the same type that occurred here some 2,000 years ago. Check at the visitor center for the schedules of conducted walks and campfire programs. You also can get information here about two self-guiding nature trails and the park's Loop Drive (see page 59).

Activities and Evening Programs. In summer, ranger-guided walks and other programs give visitors an intimate look at various aspects of the park. Program schedules vary; we suggest that you contact the park for current information prior to arrival. Several sites have been designed to make it easy to see the park on your own. The visitor center is a good place to stop and plan your visit. Evening programs may find you wanting a sweater or light jacket to ward off the chill, despite the hot summer days. These programs explore such topics as the park's wildlife and its survival, the powers of nature, and this landscape's volcanic origins. Some programs are illustrated with slides or movies and take place in the amphitheater.

Self-guiding Trails. Explore three representative areas of the park on self-guiding nature trails. **Devils Orchard Trail** helps you understand the complex environmental concerns facing Craters of the Moon. A pamphlet available at the trailhead discusses the major impacts visitors, neighbors, and managers have on the fragile lava landscape. You can walk this trail in about 20 minutes. **North Crater Flow Trail** takes you through a lava flow that includes rafted blocks (crater wall fragments) and other interesting features characteristic of basaltic lava flows, which are explained by wayside exhibits. This trail goes through one of the most recent lava flows in the park. The shiny lava flows made early explorers think the volcanic eruptions had happened only a few years before. Please stay on trails in this vary fragile area. The park was established to provide protection for its unusual landscape features. These require continuing protection and you can help provide it. **Caves Trail** allows you the opportunity to explore a lava tube. These caves formed when the surface of lava flow cooled and hardened while the interior remained molten and continued to drain. After the lava drained away, a hollow tube remained. A pamphlet at the trailhead provides a map of the cave area and tells you what to expect as you explore these lava tubes on your own. Wayside exhibits point out the most interesting lava formations along the trail. To see only Indian Tunnel, the largest of the lava tubes, will require nearly one hour.

CRATERS OF THE MOON NATIONAL MONUMENT

Wilderness area

8 Point of interest

▲ Campground

Trail

To Arco

20 26 93

PIONEER
MOUNTAINS

Visitor Center

HIGHWAY FLOW

GREEN

DRAGON

FLOWS

SERRATE FLOW

Sunset Cone
6410ft
1954m

1

▲

North Crater
Flow Trail

2

Grassy Cone
6315ft
1925m

North Crater
6244ft
1903m

Paisley Cone
6107ft
1861m

Devils Orchard
Nature Trail

3

7

Beauty Cave
Boy Scout
Cave

Indian Tunnel

Dewdrop
Cave

BLUE

DRAGON

FLOWS

North Crater Trail

Silent Cone
6357ft
1938m

BIG
CRATERS

4

Inferno Cone
6181ft
1884m

5

Snow Cone
Spatter Cones

GREAT

LAVA
CASCADES

BIG SINK

RIFT

Half Cone
6055ft
1846m

BROKEN TOP FLOW

Broken Top
6058ft
1846m

ZONE

6

Big Cinder Butte
6515ft
1986m

BLUE DRAGON FLOWS

To Carey

North

Take The Driving Tour

You can see most of the features for which Craters of the Moon is famous by a combined auto and foot tour along the Loop Drive. With several short walks included, you can make the drive in about two hours. Numbered stops are keyed to the map in the park folder.

1. Visitor Center. The 7-mile Loop Drive begins at the visitor center. Most of the drive is one-way. Spur roads and trailheads enable you to explore this lava field even further.

2. North Crater Flow. A short foot trail crosses the North Crater Flow to a group of crater wall fragments transported by lava flows. This is one of the youngest flows here. The triple twist tree and its 1,350 growth rings have in the past helped date the recency of the last flows here. Along this trail you can see fine examples of pahoehoe lava and aa lava flows (see page 26). Just beyond the North Crater Flow Trail is the North Crater Trail. This short, steep, self-guiding nature trail leads you to the vent overlooking the crater of a cinder cone.

3. Devils Orchard. Devils Orchard is a group of lava fragments that stand like islands in a sea of cinders. This marks the resting place for blocks of material from the walls of North Crater that broke free and were rafted here on lava flows. The short spur road leads to a self-guiding trail through these weird features. You can easily walk the trail in about 20 minutes. An early morning or evening visit may allow you to observe park wildlife. In springtime, the wildflower displays in the cinder gardens are glorious. In June and early July, dwarf blooming monkeyflowers give the ground a magenta cast.

4. Inferno Cone Viewpoint. From the viewpoint atop Inferno Cone, a landscape of volcanic cinder cones spreads before you to the distant mountain ranges beyond. Notice that the cooler, moister northern slopes of the cones bear noticeably more vegetation than the drier southern slopes, which receive the brunt of sunshine. If you take the short, steep walk to the summit of Inferno Cone, you can easily recognize the chain of cinder cones that defines the Great Rift. Perhaps nowhere else in the park is it so easy to visualize how the volcanic activity broke out along this great fissure in the Earth. Towering in the distance above the lava plain is Big Cinder Butte, one of the world's largest, purely basaltic, cinder cones.

5. Big Craters and Spatter Cones Area. Spatter cones formed along the Great Rift fissure where clots of pasty lava stuck together when they fell. The materials and forces of these eruptions originated at depths of approximately 40 miles within the Earth. A short, steep walk to the top of Big Craters offers a view of a series of volcanic vents.

6. Trails to Tree Molds and the Wilderness Area. A spur road just beyond Inferno Cone takes you to trails to the Tree Molds Area and the Craters of the Moon Wilderness. Tree molds formed where molten lava flows encased trees and then hardened (see page 27). The cylindrical molds that remained after the wood burned and rotted away range from a few inches to more than 1 foot in diameter.

7. Cave Area. At this final stop on the Loop Drive, a ½-mile walk takes you to the lava tubes. Here you can see Dewdrop, Boy Scout, and Beauty Caves and the Indian Tunnel. (For how these lava tubes formed, see page 30.) Carry a flashlight in all caves.

Camping and Backcountry Use

The campground has about 50 sites. These are available on a first-come, first-served basis. Reservations are not accepted. A daily fee for camping is charged. Water and restrooms are provided in the campground, but there are no showers, dump station, or hookups. Wood fires are prohibited in the park, but grills at each campsite may be used for charcoal fires. The campground accommodates both RVs and tents. During the summer, park rangers present evening programs at the campground amphitheater.

Backcountry Use. Some of the park's most intriguing landscapes lie beyond the road's end in the 68-square-mile Craters of the Moon Wilderness Area. Only two trails penetrate the wilderness, and these for only short distances. After the three-mile trail to Echo Crater runs out, you are on your own. For further exploration, you can simply follow the Great Rift and its chain of cinder cones. These landmarks help you find your way.

To explore farther afield, you should have a good topographic map and basic map skills. You can purchase such a map at the visitor center. All hikers who plan to stay overnight in the wilderness are required to register with a park ranger. Backcountry use permits are available free at the visitor center.

Each hiker should carry at least one gallon of water for each day out; even more may be necessary during the hot summer. There is no drinking water available in the wilderness. The best times for wilderness travel are May-June and September-October. Daytime temperatures are usually mild then, while nights are cool, but you must be prepared for inclement and very cold weather in these transitional months. Summer daytime temperatures climb into the 90s, and reflected heat off the lavas may be even higher. Long distance hiking is not very pleasant then, and the weight of necessary drinking water is burdensome.

Safety. Sturdy boots and long pants are necessary gear for the jagged aa lava flows. Bring clothing for both hot and cool weather; both can occur the same day in this desert climate. (See drinking water warning above.)

Regulations. Campfires are prohibited in the backcountry. Carry a self-contained backpack stove and fuel. Mechanized vehicles, including bicycles, are prohibited in the wilderness area. Pets are also prohibited in the wilderness. Pack out everything that you pack in—and any trash you find that others left behind. A good admonition is: "Take only pictures, and try not to leave so much as a footprint."

Winter Recreation

The visitor center is open every day except winter holidays. Winter hours are 8 a.m. to 4:30 p.m. Wilderness permits, topographic maps, and information are available here. To find out about current snow conditions, call (208) 527-3257. **Skiing.** Crosscountry skiing provides an enjoyable experience of the park's landscape transformed by snow. When heavy snows accumulate, usually in late November, the 7-mile Loop Drive is closed and it becomes a natural ski trail. Most of the Loop Drive follows fairly level terrain. The best months for skiing are January to March in most winters. Usually there is about 18 inches of snowpack by January and 3 feet by March. Temperatures range from 45°F to well below zero. Be prepared for inclement weather and high winds at all times. Blizzards may be encountered. **Hazards.** Skiing off the Loop Drive is allowed but not recommended. Most of the park is covered by sharp, jagged lava, and snow cover may mask cracks and caverns underneath. **Camping.** Winter camping is permitted in the main campground. The campground is not plowed; be prepared to camp in the snow. Wood fires are not permitted anywhere in the park. **Wilderness use.** The wilderness is ideal for overnight ski trips. You should be well equipped and experienced at winter camping, however. A free wilderness use permit, available at the visitor center, is required for all overnight use outside the park campground.

Both backpackers and cross-country skiers find solitude in their respective seasons in the park. Others may prefer ranger-led explorations of the park's many unusual features.

Regulations and Safety

Many management concerns, regulations, and safety tips are given under specific subjects in this handbook. Here are some other things to consider.

Precautions must be taken when you explore the park because of the rugged terrain, heat, and lack of naturally available drinking water. You will need sturdy boots, a hat, and ample, leakproof water containers. Make sure containers are watertight **before** you leave home. Exploring caves requires flashlights.

Camp only in the park campground. All other overnight use, even in winter, requires a wilderness use permit. A day-use permit is required to visit the area of the park that lies north of Highway 20-26-93.

Pets. Pets are allowed only in the campground and on the Loop Drive, but they must be kept on a leash at all times. Pets are prohibited in all public buildings, on trails, or in off-road areas.

Vehicles. All motor vehicles and bicycles must stay on paved roads only. They are not allowed on trails.

Firearms. Firearm restrictions are enforced: No hunting is allowed in the park.

Collecting. The collection, removal, or disturbance of any natural features within the park is strictly prohibited.

For contemporary explorers the driving tour and its associated trails make the safest trek routes. Exercise great caution—and close oversight of young children—at all times on your park expeditions.

Nearby Attractions

Yellowstone National Park is world famous for its geysers and mudpots, canyons and waterfalls, and wildlife and wilderness. For information write or call, Superintendent, Yellowstone National Park, WY 82190, (307) 344-7381.

Grand Teton National Park features the spectacularly scenic Teton Range and lovely lakes at its base. **John D. Rockefeller, Jr., Memorial Parkway** joins Grand Teton with Yellowstone. For information write or call, Superintendent, Grand Teton National Park, P.O. Drawer 170, Moose, WY 83012, (307) 733-2880.

Nez Perce National Historical Park includes 24 widely scattered sites in north-central Idaho that present the history of this ancestral homeland of the Nez Perce tribe. For information write or call, Superintendent, Nez Perce National Historical Park, P.O. Box 93, Spalding, ID 83551, (208) 843-2261.

Hagerman Fossil Beds National Monument, authorized in 1988, preserves Pliocene fossil sites along Idaho's Snake River. The National Park Service is planning for future needs. Facilities have not been developed. For information write or call, Superintendent, Hagerman Fossil Beds National Monument, P.O. Box 570, Hagerman, ID 83332, (208) 837-4793.

City of Rocks National Reserve is a fascinating landscape of monoliths, spires, and domes used historically by Northern Shoshone Indians and emigrants on the California Trail. It has become a mecca for recreational rock climbers. Primitive facilities. For information write, Manager, City of Rocks National Reserve, P.O. Box 169, Almo, ID 83312.

Minerva Terrace, Yellowstone

Grand Teton in winter

A Nez Perce today

Armchair Explorations

The nonprofit Craters of the Moon Natural History Association sells books, maps, and other publications at the visitor center or by mail. For a free list write to the park address on page 55. The following selected books may also be of interest.

Belknap, William J. "Man on the Moon in Idaho," *National Geographic Magazine,* Volume 119 (October, 1960).

Bonnichsen, Bill and Roy M. Breckenridge et al. *Cenozoic Geology of Idaho,* Idaho Geologic Survey, University of Idaho, 1982.

Bullard, Fred M. *Volcanoes of the Earth,* University of Texas Press, 1976.

Chronic, Halka. *Pages of Stone: The Geologic Story of Our Western Parks and Monuments,* The Mountaineers, 1984.

Clark, David R. *Craters of the Moon— Idaho's Unearthly Landscape,* Craters of the Moon Natural History Association, 1990.

Henderson, Paul A. *Around the Loop: Craters of the Moon,* Craters of the Moon Natural History Association, 1986.

Limbert, Robert W. "Among Craters of the Moon," *National Geographic Magazine,* Volume 45 (March, 1924).

McKee, Bates. *Cascadia,* McGraw-Hill, 1972.

Moser, Don. *The Snake River Country,* Time-Life Books, 1974.

National Aeronautics and Space Administration (NASA). *Volcanism of the Eastern Snake River Plain, Idaho: A*

Comparative Planetary Geology Guidebook, Washington, D.C., 1977.

Schwartz, Susan. *Nature in the Northwest,* Prentice-Hall, 1983.

Other National Park Handbooks in this series. You might enjoy other official National Park Handbooks about areas in Idaho, Wyoming, and Montana. These handbooks include: Grand Teton National Park; Nez Perce National Historical Park; Devils Tower National Monument; and Fort Laramie National Historic Site.

These informative handbooks are available at the parks or by mail from: Superintendent of Documents, U.S. Government Printing Office, Washington, DC 20402. For a list of handbooks write to: **National Park Service, Office of Information, P.O. Box 37127, Washington, DC 20013-7127.**

☆GPO: 2000—462-115/20501   Reprint 2000